SpringerBriefs in Food, Health, and Nutrition

T0183768

Editor-in-Chief

Richard W. Hartel, University of Wisconsin—Madison, U

Associate Editors

J. Peter Clark, Consultant to the Process Industries, USA
John W. Finley, Louisiana State University, USA
David Rodriguez-Lazaro, ITACyL, Spain
David Topping, CSIRO, Australia

For further volumes:
http://www.springer.com/series/10203

SpringerBriefs in Food, Health, and Nutrition present concise summaries of cutting edge research and practical applications across a wide range of topics related to the field of food science.

Alper Gueven · Zeynep Hicsasmaz

Pore Structure in Food

Simulation, Measurement and Applications

 Springer

Alper Gueven
Faculty of Engineering
Department of Food Engineering
Tunceli University
Tunceli
Turkey

Zeynep Hicsasmaz
Faculty of Engineering and Architecture
Department of Food Engineering
Trakya University
Edirne
Turkey

ISBN 978-1-4614-7353-4   ISBN 978-1-4614-7354-1   (eBook)
DOI 10.1007/978-1-4614-7354-1
Springer New York Heidelberg Dordrecht London

Library of Congress Control Number: 2013937603

Printed on acid-free paper

Springer is part of Springer Science+Business Media (www.springer.com)

Contents

Abstract

Influence of microstructure on process rates and product texture is widely accepted. Simulation of the pore structure of various food materials and numerous food processes has attracted recent attention in the food literature. Several approaches have been applied to predict transport phenomena in food materials. The first one is the continuum approach in which all the microscopic complexities are lumped into effective diffusivities in which diffusivities are empirical constants. Simulation of the characteristic pore structure depends on accurate quantitative data on the pore size distribution. Qualitative three-dimensional (3-D) imaging techniques such as X-ray micro-tomography and magnetic resonance imaging (MRI), and two-dimensional (2-D) imaging techniques such as scanning electron microscopy (SEM) and transmission electron microscopy are used to analyze the cell structure. X-ray microtomography and MRI are 3-D non-invasive techniques that allow reconstruction of the pore structure, but the sophisticated procedure complicates combining the restructured model with transport equations. Mercury porosimetry and liquid extrusion porosimetry are used for quantitative evaluation of the cell structure. Drawback of porosimetry is: Laplace-Young equation of capillary flow dictates that the intruded/extruded pore diameter is inversely proportional with capillary pressure. Thus, cell size distribution from intrusion/extrusion data is erroneous since the assumption that pore sizes regularly increase/decrease as intrusion/extrusion advances is not physically true. A mathematical model that simulates experimental intrusion/extrusion curves by randomly distributing different cell sizes within a geometric unit cell remedies this drawback. Such a model is called a geometric network model, and is the second approach that can be used to simulate the pore structure and transport phenomena through the porous medium. Geometric network models used in petroleum reservoir engineering, soil science, and catalysis are for relatively low porosity materials (porosity ≤ 0.4). Food materials generally have high porosity (0.5–0.9), thus direct application of geometric network models to foods is difficult. There are very few applications of geometric network models to food materials. 1-D approaches have been applied to simulate the pore structure of food materials, and 2-D approaches have been applied to simulate the rehydration and reconstitution of ready-to-eat foods. Herein: The importance of the pore structure in terms of food processing and product quality is emphasized; methods of obtaining accurate data on the pore structure of food materials are described; advantages and drawbacks of several available predictive mathematical models are presented; and how the geometric network model can be used to predict the pore structure and transport through porous food materials is described.

Chapter 1
Introduction

Porosity is the ratio of the pore volume to the apparent volume of the porous medium. Food products such as baked, extruded, puffed, dried and frozen foods have an inherent porous micro-structure that gives the product its characteristic texture (Roca et al. 2006; Gogoi et al. 2000) measured in terms of physical properties such as tensile strength, compressive strength and stiffness (Stasiak and Jamroz 2009). Porosity is the macroscopic pore structure parameter which develops in conjunction with the microscopic pore structure. Microscopic pore structure parameters can be summarized as the specific surface area, average pore size and pore size distributions based on volume, surface area and number of pores, cell wall thickness, pore shape distributions, polydispersity indices for pore sizes and shapes and the pore interconnectivity. These microscopic pore structure parameters also affect transport properties such as thermal diffusivity and moisture diffusivity as well as texture.

1.1 Importance of Pore Structure on the Quality of Convenience Foods

One of the most striking features of the last decade has been convenience foods due to the lack of time and energy and decrease in ability to prepare food. Convenience foods refer to the easy to prepare dried/fried foods, baked/extruded/puffed snacks and composite foods that consist of compartments of various textures. Change in porosity and the cellular structure of food materials during processing (Madiouli et al. 2012; Khalloufi et al. 2009; Hussain et al. 2002), as in the case of convenience foods, are known to influence processing rates, mechanisms, and thus the product quality. Porosity and the cellular structure influence the final texture and mouth feel of convenience foods (Prakotmak et al. 2010).

Quality of dried ready-to-eat foods depends on reconstitution upon rehydration. Formation of the characteristic cell structure upon drying influences rehydration ability, and thus the final quality. Rehydration ability of dried foods depends on

A. Gueven and Z. Hicsasmaz, *Pore Structure in Food*,
SpringerBriefs in Food, Health, and Nutrition, DOI: 10.1007/978-1-4614-7354-1_1,
© Alper Gueven and Zeynep Hicsasmaz 2013

liquid uptake and solids leaching of the food matrix. Rehydration studies are based on empirical (Troygot et al. 2011) and semi-empirical approaches (Saguy et al. 2005) as well as mechanistic models such as Fick's Second Law of Diffusion (Troygot et al. 2011). Imaging based techniques proved rehydration to be a non-Fickian process driven by capillary penetration. Thus, the pore network model offers improvement on predicting rehydration ability (Prakotmak et al. 2010) as long as the porosity and pore size distribution of the food material are known.

Deep-fat frying is another way of manufacturing convenience foods. Deep-fat frying can be defined as a combined cooking-drying process in which air and moisture in the food matrix is replaced by oil. Therefore, the quality of deep-fried foods is also closely related with capillary flow in porous media.

1.2 Influence of Pore Structure on Moisture Migration

Moisture migration is known to be one of the major causes of loss in crispiness and decrease in the shelf life of composite foods. The mechanism involves surface sorption followed by migration of moisture into the porous food product. Moisture sorption is related with the chemical composition of the food surface, whereas moisture diffusivity is a more complicated phenomenon. The main transport mechanism in moisture migration is accepted to be moisture diffusion described by Fick's Second Law of Diffusion. Moisture diffusivity is a function of temperature and the moisture content of the food material. However, studies on moisture diffusion suggested a strong dependence of effective moisture diffusivity on the microscopic structure of the food material (Prakotmak et al. 2010). Increase in porosity accompanied by an open pore structure is claimed to increase moisture diffusivity to a great extent (Baik and Marcotte 2002).

1.3 Change in Pore Structure During Food Processing

Food processing operations such as drying puffing, extrusion, freezing result in changes in porosity, and thus the micro-structure. Variation of porosity during processing depends on shrinkage and/or collapse (Rahman 2003; Khalloufi et al. 2009, 2010) which are two opposing phenomena. Collapse causes decrease in the pore volume, while shrinkage increases the pore volume. Thus, mathematical modeling of porosity and its changes during processing has attracted attention.

Baking is another food processing operation during which the physical state of dough changes completely and is converted into spongy products with different mechanical properties. Dough passes through complex physical and chemical changes during baking. Changes in structure occur mainly through evaporation of water from the crumb, also termed as vapor-induced puffing, accompanied by protein denaturation and starch gelatinization. Control of the temperature and humidity is very important in terms of the desired expansion volume, pore inter-connectivity, and a crust envelope with desired characteristics. Desired crumb structure and expansion volume depends on the initial stages of crust formation that requires control of the surface humidity (Ahrné et al. 2007) which improves extensibility of the surface promoting heat transfer into the crumb (Le-Bail et al. 2011; Kocer et al. 2007; Hicsasmaz et al. 2003). Moisture transport from the crust to the crumb is a complex diffusion problem in which moisture travels through a capillary network under the effect of capillary pressure and a water vapor con-centration gradient (Luyten et al. 2004). It is also very difficult to retain a crispy crust during the cooling stage after baking. Temperature and moisture gradient between the crumb and the crust, and also at the interface between the crust and the environment leads to moisture transport into the crust causing a loss in crispiness. The rate of moisture transport is influenced by the cell size distribution, interconnectivity of the cells and the cell wall thicknesses (Primo-Martin et al. 2010). Therefore, control of moisture diffusion through the crumb by monitoring formation of the crumb structure is essential in obtaining baked products with uniform structure (Altamirano-Fortoul et al. 2012). A mathematical model that relates monitored changes in the cell structure to the transport of moisture through the continuously changing capillary structure would be useful in controlling the oven conditions to manufacture products with uniform microstructure.

1.4 Effect of Pore Structure on Transport Properties

Transport properties such as thermal diffusivity and mass diffusivity depend on the changing and/or developing pore structure. Although moisture diffusivity and its modeling in the porous food system received considerable attention, thermal diffusivity has not been studied equally well despite its importance especially during freezing and drying (Carson et al. 2004; Sablani and Rahman 2003). This is partly due to the lack of experimental data, and also due to the complexity of the system that does not allow application of simple additive models.

Thermal conductivity is an important physical property used to estimate the rate of conduction in food processing operations, such as drying, freezing and freeze drying. Thermal conductivity is the transport property that affects the rate of heat conduction. Composition and heterogeneity of the food affects thermal conduc-tivity (Hamdami et al. 2003), and usually an effective thermal conductivity or thermal diffusivity is defined for porous foods. Thermal conductivity predictions of porous foods involve a high level of uncertainty. This is because the thermal conductivity of air is an order of magnitude less than the food components, thus allowing a wide range of theoretically possible effective thermal conductivity values for a given porosity when simple additive models are used.

Thermal conductivity during freezing and freeze drying is even more complex, since it is not only influenced by the heterogeneous structure of the food, but it is also affected by latent heat transfer accompanied by vapor diffusion. Water that evaporates at the high temperature side diffuses in the pore space according to the vapor pressure gradient caused by the temperature gradient. Moisture then condenses at the low temperature side, transporting latent heat through the pores.

Measurement of thermal conductivity of food materials at freezing temperatures and a few degrees below is subject to high variability due to a moving ice front. Under such conditions, use of a predictive model, more precise than measurements, becomes preferable. Integration of the pore network model to the well-known parallel, perpendicular, dispersed phase and Krischer models of thermal conductivity can improve thermal conductivity predictions.

1.5 Effect of Pore Structure on Texture

Mechanical properties of cellular food materials are governed jointly by their cell wall material property and their pore structure. Strong correlations have been reported between porosity and mechanical properties such as stiffness, loss modulus and shear strength for a variety of food materials (Gogoi et al. 2000). The general approach has been to use the empirical power law relationship between porosity and mechanical strength (Fang and Hanna 2000) or to explain the relationship by the Gibson & Ashby model (Liu and Scanlon 2003). However, there are frequent reports in the literature that either the relationship between porosity and the mechanical strength does not fit power law, or the Ashby exponent value does not explain the relationship. Neither model takes the micro-structural features of the cellular samples into consideration which can be different although porosity may be the same. Another problem with cellular manufactured food materials is their morphological anisotropy (Li et al. 2006) and low polydispersity index (Kocer et al. 2007).

Experimental determination of the mechanical properties of porous foods is a challenge because of high sample-to-sample variability owing to random variations in microstructure. In such cases, predictive texture models may provide relevant information on the reasons of experimental variability in mechanical properties. Conversely, predictive modeling can also provide information on the basic features of the microstructure that will provide a desired set of mechanical properties. Integration of a pore network model with a texture model that utilizes finite element techniques (Guessasma et al. 2011) can be used to predict micro-structure-texture relationships.

1.6 Mathematical Pore Structure Models

In recent years, food science and engineering have been experiencing an encouraging transition from empirically based to physically based models. The continuum approach and the pore network model are the two basic approaches in the case of porous food materials. In the continuum approach, the entire pore space is considered as macroscopically consistent with the appearance of the whole. Thus, all microscopic complexities such as cell size and shape, connectivity of the pores and cell wall thickness are lumped into effective diffusivities (Prakotmak et al. 2010).

The second approach is based on a discrete model which is the pore network model. This type of model integrates micro-structural factors by mapping the internal structure of the porous material as a network of randomly placed small-large and interconnected-noninterconnected pores. Interconnectivity can be accounted for in a one-dimensional (1-D), two-dimensional (2-D) or three-dimensional (3-D) lattice. This approach is widely used in petroleum reservoir engineering, soil science and catalysis, which involve materials with relatively lower porosities (porosity ≤ 0.4) when compared with the food materials (porosity ≥ 0.4).

Chapter 2
Quantitative Measurement of the Pore Structure

This is a special area of research in which there are numerous techniques, equipment and software that allow determination of the pore structure with respect to the number of pores, pore size and shape distributions (Kocer et al. 2007; Hicsasmaz et al. 2003), number of cell faces, cell wall thickness, pore interconnectivity and polydispersity index (Trater et al. 2005). Pore structure analysis is a high technology field in which microscopic techniques such as light microscopy, scanning electron microscopy (SEM) and transmission electron microscopy (TEM) besides numerous image analysis techniques primarily used to visualize the human body for purposes of medical diagnosis are well-adapted. X-ray microtomography (Frisullo et al. 2012; Trater et al. 2005) is one such technique in which 2-D sections can be imaged non-destructively, and then a qualitative 3-D reconstruction can be performed. 2-D micro-tomography images and related software allow measurement of cell wall thicknesses with accuracy. Magnetic resonance imaging (MRI) (Wagner et al. 2008; Ishida et al. 2001), SEM and environmental scanning microscopy (ESEM) (Stokes and Donald 2000) also provide data on the pore structure, SEM being one of the most widely used methods. Ultrasound (Lagrain et al. 2006) is used in microstructure measurements. Mercury porosimetry (Hicsasmaz and Clayton 1992) and liquid extrusion porosimetry (Datta et al. 2007) are quantitative methods which do not use imaging techniques, but that use the capillary penetration technique.

2.1 Capillary Penetration Techniques

2.1.1 Mercury Intrusion Porosimetry

Mercury porosimetry is a capillary penetration technique used to predict the pore volume intruded/extruded corresponding to a certain capillary pressure. In mercury porosimetry, relatively higher capillary pressures are used due to high surface tension and contact angle of the non-wetting mercury. Detection of capillaries with

A. Gueven and Z. Hicsasmaz, *Pore Structure in Food*,
SpringerBriefs in Food, Health, and Nutrition, DOI: 10.1007/978-1-4614-7354-1_2,
© Alper Gueven and Zeynep Hicsasmaz 2013

diameters greater than 200 μm is not possible using mercury porosimetry. In mercury porosimetry, the sample cell is first evacuated, and then the pressure is increased incrementally, meanwhile monitoring the mercury filled specimen pore volume continuously upon each increment in pressure (Hicsasmaz and Clayton 1992). If capillaries with diameters larger than 200 μm occupy considerable volume, then mercury penetration is accomplished in a very short incremental pressure range and a sound structural analysis cannot be performed.

2.1.2 Liquid Extrusion Porosimetry

In liquid extrusion porosimetry, the specimen is first imbibed in a special liquid with very low surface tension, and the capillaries are emptied progressively from largest to the smallest by incremental increase in pressure, meanwhile monitoring the emptied pore volume continuously. The membrane arrangement used is such that, extruded liquid can pass through the membrane selectively at gas pressures applied within the range of the test (Datta et al. 2007).

2.2 Imaging Techniques

2.2.1 Destructive Techniques

SEM is well-suited for quantitative analysis of the pore structure, since it allows a wide range of magnification, a high depth of field, and produces digital data fit for image analysis. SEM combines the best aspects of light microscopy and TEM (James 2009). SEM is widely used for micro-structural analysis. However, analysis of many sections is required to get statistically relevant results, and the information on total pore volume and pore size distribution is not reliable due to the discontinuous nature of 2-D sections (Gropper et al. 2002; Autio and Salmenkallio-Martilla 2001). This is because 2-D sections are prepared by cutting to expose the cross-section. Cutting is a destructive technique which may in turn alter the structural features of the specimen.

Individual pores are sliced off-center and the diameters depend on the depth of the cut (Scanlon and Zghal 2001; Campbell and Mougeot 1999). Therefore, micro-structural data obtained from SEM sections are semi-quantitative. Reliable assessments on micro-structure can be made by combining the results of SEM analysis with those from capillary penetration techniques. Destructive nature of SEM images does not allow quantization of pore interconnectivity and cell wall thickness. These are in fact the most important features of the microstructure that influence mechanical strength, bulk infusibility by liquids and effective diffusivity of gases.

It is difficult to obtain adequate contrast between the air and solid phases using light microscopy and SEM. Although lighting and the angle of illumination are important, even careful arrangement proves to be inadequate for specimens with thin cell walls. Cell walls on the 2-D section can be painted to enhance contrast when light microscopy is used (Dogan et al. 2013; Kocer et al. 2007; Hicsasmaz et al. 2003). However, even careful painting can introduce extra artifacts. Therefore, these techniques provide semi-quantitative results on pore size and shape distributions and their relationship with processing parameters and ingredient mixtures.

Imaging by SEM requires special preparation of samples like food materials that contain water. SEM is well-suited for dried food samples (Bai et al. 2002) which further results in shrinkage, and thus alteration of the microstructure. The dried sample is sputtered by a metal coating to enhance conductivity, and thus to obtain high quality images. Cryo-SEM is an alternative to avoid the disadvantages of drying. In this case, cold-stage imaging is performed on frozen samples, but formation of ice crystals cause artifacts, and makes it difficult to perform quantitative analysis. ESEM (Electroscan SEM) allows imaging of hydrated samples. This is possible due to the design of ESEM which allows the electron chamber under high vacuum, while keeping the specimen chamber under moderate vacuum.

2.2.2 Non-invasive Techniques

X-ray microtomography is a non-invasive technique allowing visualization of the internal structure of a sample based on the local variation of the X-ray attenuation coefficient. X-ray microtomography allows scanning of the whole sample. This technique is quite new in the field of food engineering (Leonard et al. 2008; Babin et al. 2006; Haedelt et al. 2005; Trater et al. 2005; Lim and Barigou 2004; van Dalen et al. 2003). The general methodology involves targeting the specimen with a polychromatic X-ray beam. X-rays transmitted by the specimen fall on specially-designed X-ray scintillators that produce visible light, which is recorded by a charge-coupled device camera. Transmitted intensity is related to the integral of the X-ray attenuation coefficient (μ) along the path of the beam according to the Lambert–Beer's law. The X-ray attenuation coefficient depends on the density (ρ) and atomic number (z) of the material and on the energy of the incident beam (E) given by:

$$\mu = \rho\left(a + \frac{bz^{3.8}}{E^{3.2}}\right) \tag{2.1}$$

where a and b are constants (Leonard et al. 2008).

A tomographic scan is accomplished by rotating the specimen between 0 and $180°$ about an axis perpendicular to the X-ray beam while collecting radiographs of the specimen at small angular increments. The radiographs are then reconstructed into a series of 2-D slices using the back-projection software (Brun et al. 2010) that maps estimates of attenuation coefficients, which depend on density variations within the specimen. The series of slices, covering the entire sample, can be reconstructed into a 3-D image that can either be presented as a whole or as virtual slices of the sample at different depths and in different directions. Reconstruction of cross-sections at depth increments as low as 1 μm, and along any desired orientation is possible using special software (Brun et al. 2010). The resulting 3-D and/or 2-D reconstructed images are formed of two grey levels corresponding to the two phases of the porous material. The low grey levels (dark voxels) correspond to the pore space, and the high grey levels (bright voxels) correspond to the solid cell walls.

MRI is ideally suited for non-invasive imaging of water, and has been applied to food materials (Horigane et al. 2006; Troutman et al. 2001; Cornillon and Salim 2000). MRI imaging is not a quantitative method of microstructure measurement, but it is very powerful in following moisture migration problems. The effect of micro-structure on moisture migration can further be studied by relating MRI images with X-ray microtomography images (Weglaz et al. 2008). MRI is based on nuclear magnetic resonance (NMR) signals sampled in the presence of magnetic fields to obtain spatial resolution depending on the NMR properties of the specimen (proton density, T1 and T2). Proton density is the NMR property of the specimen that originates from the protons of carbohydrates, proteins and fats in the solid matrix. Differences in T2 decay signals describe the water of hydration, free water and the lipids. Short T2 decays correspond to the water of hydration, while long T2 decays correspond to free water. Time domain signals sampled under the influence of magnetic field gradients are converted into 1-D, 2-D or 3-D images called the "k-space". The MR image is obtained after a series of complex Fourier transforms. The key issue in the MRI technique is to collect enough data points in the "k-space", i.e. resolution in time and space that can describe the specimen (Weglarz et al. 2008).

2.2.3 *Quantitative Image Analysis*

Measuring the pore structure parameters such as the total pore volume, pore size distributions, polydispersity index, pore shape factors, interconnectivity of the pores and cell wall thicknesses requires the use of special image analysis software regardless of whether the images are obtained by destructive or non-invasive techniques. Image analysis software is used to perform an action called tresholding on the images to convert grey level images to binary images. The software automatically determines the treshold grey tone by maximizing the entropy of the histogram, s, given by (Leonard et al. 2008; Kocer et al. 2007; Hicsasmaz et al. 2003):

$$s = - \sum p_i \log(p_i) \tag{2.2}$$

where p_i is the probability of any pixel grey value in the image. Then, the value of 1 is assigned to all pixels whose intensity is below the threshold grey tone, and the value of 0 is assigned to all the others. An average threshold grey value among the analyzed 2-D sections needs to be defined to carry out quantitative image analysis on 3-D reconstructions. Tresholding results in a binary image regardless of whether quantitative image analysis is carried out on 2-D sections or 3-D reconstructions. The binary image is further filtered to remove the black spots (Leonard et al. 2008; Kocer et al. 2007; Hicsasmaz et al. 2003). Colored parts of the binary image correspond to the cell walls, black parts correspond to the pores. Then, the pore structure parameters are evaluated depending on the capabilities of the image analysis software.

Image analysis software should have special capabilities to increase contrast between the gas and solid phases before quantitative analysis of images obtained by 2-D destructive techniques is performed. Image analysis software use techniques such as image subtraction and edge detection that involve complex mathematical methods such as Fourier transforms. However, these enhancement methods can also introduce extra artifacts. Objective quantitative measurement of the pore structure parameters is not possible using 2-D images obtained from 2-D destructive techniques as a result of the contrast problem, and high section to section variability.

X-ray microtomography is a powerful tool for quantitative analysis of the pore structure. Trater et al. (2005) defined a representative volume element and suggested a method to calculate various pore structure parameters. Defined representative volume element was divided into a number of equidistant 2-D slices. The pore volume (V_i) of the selected representative volume element was found as;

$$V_i = t \left(\sum_{n=1}^{N_i-1} \frac{A_{v,n,i} + A_{v,n+1,i}}{2} \right) + \frac{t}{3} \left(A_{v,1,i} + A_{v,n_i,i} \right) \tag{2.3}$$

where $A_{v,n,i}$ is the pore area on slice n, t is the constant spacing between any two slices, and N_i is the number of slices that pore i appears. Diameters of the pores (D_i) in the representative volume element were calculated assuming spherical geometry:

$$D_i = 2 \left(\frac{3V_i}{4\pi} \right)^{1/3} \tag{2.4}$$

The average pore diameter (D_{mean}) was calculated as the mean diameter of all the pores within the representative volume element:

$$D_{mean} = \frac{\sum_{i=1}^{M} D_i}{M} \tag{2.5}$$

where M is the number of pores in the representative volume element. The weighted average pore diameter (D_{wt}) was calculated as:

$$D_{wt} = \frac{\sum_{i=1}^{M} D_i V_i}{\sum_{i=1}^{M} V_i} \qquad (2.6)$$

where the individual pore volumes (V_i) are the weighting factors. Porosity of the representative volume element (ε) was calculated as:

$$\varepsilon = \frac{\sum_{n=1}^{T} \sum_{i=1}^{M_n} A_{v,n,i}}{\sum_{n=1}^{T} \sum_{i=1}^{M_n} A_{v,n,i} + \sum_{n=1}^{P} A_{s,n}} \qquad (2.7)$$

where $A_{v,n,i}$ is the pore area on slice n, M_n is the total number of pores on a slice, $A_{s,n}$ is the total solid area on a slice and T is the number of slices in a representative volume element. The average cell wall thickness (t_{wall}) was evaluated as:

$$t_{wall} = \frac{\sum_{n=1}^{R} A_{s,n} / \sum_{i=1}^{M_n} P_{n,i}}{R} \qquad (2.8)$$

where $P_{n,i}$ is the total perimeter of pore i in slice n. Fractional cell interconnectivity (φ) was calculated as:

$$\varphi = \frac{\sum_{i=1}^{M} \left(\frac{\sum_{i=1}^{N_i} l_{n,i}}{\sum_{i=1}^{N_i} P_{n,i}} \right)}{M} \qquad (2.9)$$

where $l_{n,i}$ is the total length of line segments needed to close the discontinuous portions on each pore i in slice n, $P_{n,i}$ is the total perimeter of pore i in slice n, N_i is the number of slices that pore i appears and M is the number of pores in the representative volume element. Weighted fractional cell interconnectivity was evaluated as:

$$\varphi_{wt} = \frac{\sum_{i=1}^{M} \left(\frac{\sum_{i=1}^{N_i} l_{n,i}}{\sum_{i=1}^{N_i} P_{n,i}} \right) V_i}{\sum_{i=1}^{M} V_i} \qquad (2.10)$$

where the pore volumes (V_i) are the weight factors. Polydispersity index (I) was calculated as

$$I = \frac{\sum_{r=1}^{R} D_r N_r / \sum_{r=1}^{R} N_r}{\sum_{r=1}^{R} D_r^2 N_r / \sum_{r=1}^{R} D_r N_r} \qquad (2.11)$$

2.3 Applications of Pore Structure Analysis to Food Materials

Quantitative pore structure analysis has been widely used to evaluate the microstructure of porous foods and lately 3-D imaging techniques are applied to study the pore structure of foods, and to evaluate the effect of processing on structure-related characteristics of food materials.

Drying studies followed by tomographic image analysis showed that increase in the drying temperature resulted in an increase in porosity due to the development of higher pressure gradients within the sample resulting in higher rates of moisture loss (Leonard et al. 2008). This semi-quantitative approach can be used to compare the pore structures developed during various drying techniques. Continuous monitoring of the pore structure coupled to a pore network model can be used to control drying so as to manufacture dried food products with the optimum pore structure characteristics.

Baked and extruded foods are characterized by pore sizes within a few orders of magnitude (Dogan et al. 2013; Kocer et al. 2007; Hicsasmaz et al. 2003). Thus, it is very difficult to obtain objective quantitative results. Large range of pore sizes result in large specimen-to-specimen variations requiring a vast number of 2-D sections to be analyzed. Non-invasive image acquisition offers vast opportunities for objective quantitative evaluation of baked and extruded food materials.

Images of baked and extruded foods obtained by light microscopy (Kocer et al. 2007; Hicsasmaz et al. 2003) and X-ray microtomography (Trater et al. 2005) showed a bimodal distribution of pore sizes in which larger pores were located towards the interior, while relatively smaller pores were abundant towards the crust. This is due to the faster cooling of the crust since the temperature gradient at the interface with the ambient is very high leading to high convective heat transfer rates, thus faster cooling and setting in the crust region.

Similar behavior was observed in Supercritical Fluid Extrusion(SCFX) extrudates with practically a non-porous outside layer, a large number of small cells towards the surface, and a small number of large cells towards the center. SCFX is a process in which supercritical CO_2 (SC-CO_2) is used as the blowing agent instead of steam (Cho and Rizvi 2009). CO_2 loss mainly happens from the surface and a high concentration gradient at the interface with the ambient causes the rate of CO_2 loss to increase. After the initial stage of CO_2 loss the SCFX surface acts like a seal, since diffusion distance through the surface increases due to a large number of small cells. This further allows the CO_2 bubbles towards the center to increase in size forming fewer pores with larger size. SC-CO_2 forms gas bubbles at the die exit upon expansion to the ambient. Bubbles towards the center expand freely, whereas the nuclei at the surface are restricted by the die surface. Besides, nuclei towards the center have more time to expand (Potente et al. 2006). Finally, SCFX extrudates require post-processing treatment, since SCFX is a low temperature processes that does not allow structure setting. Setting occurs during drying upon simultaneous heat and mass transfer. The surface sets first, limiting bubble growth at the surface.

Images of 2-D slices of SCFX-extrudates obtained by X-ray microtomography showed that the pores are elongated in the longitudinal direction, whereas they are spherical in the radial direction (Cho and Rizvi 2009). This finding is also true for steam extrudates (Karkle et al. 2012; Robin et al. 2010; Trater et al. 2005), and probably occurs due to the orientation of the shear field as the screw pushes the material through the barrel towards the die and out of the die to the ambient resulting in elastic recovery (Moraru and Kokini 2003; Pai et al. 2009). Image analysis on 3-D reconstructed images of SCFX extrudates showed that increase in the volumetric flow rate of the melt due to an increase in SC-CO_2 level caused this type of anisotropy to increase. Anisotropy diminished with addition of fibrous materials into the feed mix (Karkle et al. 2012; Parada et al. 2011). This is because elastic recovery at the die exit reduces, and alignment of the fibers in the direction flow hinders radial extension.

Porosity of baked and extruded products increases and their bulk density decreases with increasing expansion (Dogan et al. 2013; Kocer et al. 2007; Trater et al. 2005; Hicsasmaz et al. 2003). More expanded products are characterized by large pore sizes and lower polydispersity indeces (Kocer et al. 2007; Trater et al. 2005) due to the coalescence of the gas bubbles promoting interconnectivity. 3-D image analysis by Trater et al. (2005) showed that coalesce of gas bubbles upon expansion causes a decrease in the number of pores. Addition of fiber into the feed mixture caused a larger number of cells, but with similar expansion characteristics (Karkle et al. 2012; Parada et al. 2011). Hindered radial expansion by fiber alignment favored more nucleation sites (Chanvrier et al. 2007) increasing the number of cells, but decreasing cell diameters as observed from 2-D sections of X-ray microtomography (Karkle et al. 2012). On the other hand, both the number of pores and the pore size increased by addition of proteins in the extrusion feed mix (Mesa et al. 2009; Cheng et al. 2007). Protein–protein interactions strengthen cell walls supporting expansion, while foaming properties of proteins lead to more nucleation sites for water vapor increasing the number of pores.

Quantitative analysis of X-ray microtomography images (Trater et al. 2005) showed that the cell wall thickness increases with expansion which is contrary to the conventional logic that the cell walls should stretch and get thinner with increasing expansion. This finding is supported by the fact that cell wall thickness increases as the number of pores decrease due to the coalescence of gas bubbles during expansion. This indicates that contractile forces due to surface tension and elasticity cause the ruptured walls to be absorbed into the intact cell walls causing the cell wall thickness to increase. Quantitative image analysis of 3-D recon-structed images (Trater et al. 2005) indicated low degree of pore interconnectivity. Contrary to Trater et al. (2005), Cho and Rizvi (2009) found that cell walls become thinner as expansion increases. This is probably due to the use of different image analysis software and methodology. Measurement of microstructure on 3-D reconstructed images from X-ray microtomography depends on the capabilities of the image analysis software besides standardization and validation of the methodology.

Image analysis on 3-D reconstructed images of SCFX extrudates (Cho and Rizvi 2009) showed that increase in the volumetric flow rate of the melt due to an increase in $SC-CO_2$ level caused more pore interconnectivity. Similar results were obtained for steam extrudates indicating that increase in expansion leads to the increase in pore size (Dogan et al. 2013; Trater et al. 2005).

Chapter 3
Pore Structure and Texture

Dependence of texture on the pore structure is one of the most important fields of research. This is due to high sample-to-sample variations in application of mechanical tests. Tension, compression and flexure tests are the most common mechanical tests used in texture evaluation. Textural properties are related to microstructural features by using mechanical models, especially for brittle, solid cellular foods that have undergone glass transition during structural setting.

3.1 Mechanical Properties

Modulus of elasticity is the material property that can be obtained from a compression test. A sample of known length and cross-sectional area is compressed under a uniaxial load. The slope of the linear region of the stress–strain curve (Fig. 3.1a) is the modulus of elasticity. Compressive stress, failure stress or fracture stress is the stress at complete failure (Fig. 3.1a). The sample is buckled during compression, thus the measured stress is corrected with respect to the initial cross-sectional area as:

$$\sigma_e = \frac{F}{A_0} \tag{3.1}$$

where σ_e (N/m^2) is the engineering stress, F is the uniaxial force (N) and A_0 (m^2) is the initial cross-sectional area of the specimen perpendicular to the direction of application of the uniaxial force. Engineering strain is defined as:

$$\varepsilon_e = \frac{l - l_0}{l_0} \tag{3.2}$$

where ε_e (dimensionless) is the engineering strain, l (m) is the strain corresponding to a particular engineering stress and l_0 is the original specimen length. The linear region of the stress–strain curve that is used to calculate the compressive modulus usually occurs between strain values of 0.05 and 0.25 %.

A. Gueven and Z. Hicsasmaz, *Pore Structure in Food,*
SpringerBriefs in Food, Health, and Nutrition, DOI: 10.1007/978-1-4614-7354-1_3,
© Alper Gueven and Zeynep Hicsasmaz 2013

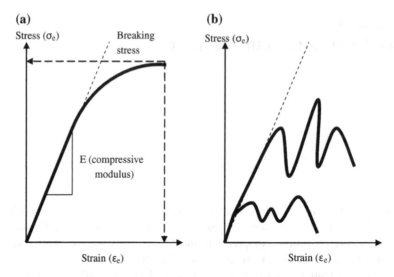

Fig. 3.1 **a** Typical stress–strain curve, **b** typical stress–strain curve for cellular foods

Brittle solid cellular foods such as extrudates produce irregular and irreproducible stress–strain curves (Fig. 3.1b). This is because materials that undergo a glass transition shatter upon very small strain. Brittleness in solid cellular foods causes cell wall failure under compression. Such a local fracture may remain as a local event, or may trigger major failure. Successive failures result in drops in stress. A jaggedness analysis can be used as a measure of crispiness of solid cellular foods (Agbisit et al. 2007). Jaggedness parameters are the number of peaks (N) on the stress–strain curve for a specimen compressed until a preset strain value. S is the area under the stress–strain curve (S) and d is the distance of puncture which can be measured from the stress–strain curve. Crispiness work (W_c) can be expressed as:

$$W_c = \frac{S}{N}$$
(3.3)

Another method used to evaluate the mechanical properties of solid cellular foods is the flexure test. The flexure test method measures behavior of materials subjected to simple beam loading. A flexure test is advantageous because tension and compression tests can be performed together in a flexure test. Tension side is the convex side, and flexural strength is defined as the maximum stress in the outermost fiber of this convex side (Fig. 3.2). On the other hand, concave side of the specimen is the compression side (Fig. 3.2). The disadvantage of the flexure test is that tension on one side, and compression on the other side of the specimen during bending causes shear in the middle. The shear effect can be minimized by keeping the specimen length to the specimen depth (Fig. 3.2) in between 16 and 64. 3-point and 4-point bending tests are the most commonly used flexure tests.

Fig. 3.2 Schematic representation of the flexure test (3-point bending test)

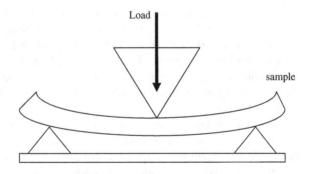

Maximum fiber stress and maximum strain are calculated for increments of load. Results are plotted in a stress–strain curve. Flexure strength is calculated at the surface of the specimen on the tension side. Flexural modulus is calculated from the slope of the stress–strain curve. The slope of the force–time data can be used to calculate the flexural modulus:

$$E_f = \left(\frac{dF}{dt}\right)\left(\frac{L^3}{4ved^3}\right) \tag{3.4}$$

where E_f (N/m^2) is the flexural modulus, dF/dt (N/s) is the slope of the force–time data, L (m) is the distance between the supports of the bending cell, v is the speed of deformation, e (m) is the thickness and d is the average cross-sectional diameter of the specimen. Breaking stress upon bending, σ_{break} (N/m^2) is calculated as:

$$\sigma_{break} = F_{break}\frac{3L}{2ed^2} \tag{3.5}$$

where F_{break} (N) is the breaking force.

3.2 Mechanical Models Describing Pore Structure-Texture Relationship

Gibson & Ashby model is widely used to describe the mechanical behavior of porous solid cellular materials. The stress–strain curve under compression is divided into three regions: elastic, collapse and densification. The behavior is mainly controlled by the relative density of the foam with respect to the solid base material. At low densities, experimental results indicate that the modulus of elasticity (E) of cellular solids is related to their density (ρ) by:

$$\frac{E}{E_{solid}} = C\left(\frac{\rho}{\rho_{solid}}\right)^n \tag{3.6}$$

where E_{solid} is the modulus of elasticity and ρ_{solid} is the density of the nonporous material, ρ/ρ_{solid} is the reduced density due to the presence of pores, C and n are microstructure-related constants inherent to the cellular solid material. C and n depend on pore interconnectivity, geometrical arrangement of the pores such as the angle of intersection, shape of the cell walls that serve as struts (curvature, cross-sectional shape and uniformity of the cell walls) and whether the material is anisotropic or not. The simplest model that can account for the microstructure is the periodically repeating cubic unit cell in which the cell walls are represented by struts and beams.

Gibson & Ashby cubic cell model follows an approach similar to the pore network model. The pore network model (Chap. 6) simulates the pore structure by geometric placement of all representative pore sizes in a cubic cell that repeats the placement geometry throughout the sample. Thus, a pore network model that describes the cellular microstructure with respect to experimentally measured porosity and pore size distribution can be integrated with the Gibson & Ashby model to implement texture-related mechanical properties of solid cellular food materials.

3.3 Solid Cellular Food Materials

Cho and Rizvi (2009) compared their experimental results on pore diameter and cell wall thickness obtained from the 3-D reconstructed images of X-ray micro-tomography with those calculated from a modified version of the Gibson & Ashby cubic model (Olurin et al. 2002). Results calculated using the cubic cell model were in accordance with those measured from the 3-D reconstructed images showing that the cubic cell model consisting of beams and struts for the cell walls gives realistic results when applied to SCFX extrudates.

Flexure modulus and breaking stress of brittle cellular foods decreased with increasing expansion (Agbisit et al. 2007; Cho and Rizvi 2009). It was found that thicker cell walls together with smaller pore diameters have higher elastic and flexure moduli (Karkle et al. 2012; Agbisit et al. 2007; Cho and Rizvi 2009) in relevance with the Gibson & Ashby cubic model. Using fiber in the extrusion feed mix strengthened the pore structure in spite of thin cell walls and smaller pore size (Karkle et al. 2012). This is because increase in the number of pores caused shorter length of the pores that decreased the beam length in the Gibson & Ashby model reinforcing the pore structure. Reinforcement of the strength of the cell walls, decrease in anisotropy, and a narrower range of pore size distribution resulted in an increase in crispiness work (Eq. 3.3). Similarly, addition of proteins to the extrusion feed mix resulted in an increase in material strength (Mesa et al. 2009).

Chapter 4
Transport in Porous Media

Processes that occur in porous media such as drying, vapor-induced puffing and rehydration take place in a dynamic environment constrained by initial porosity and/or changes in porosity as the process proceeds. In the literature, modeling of porosity is performed using fundamental concepts (Datta 2007) or empirical correlations (Mayor and Sereno 2004; Saguy et al. 2005).

Fundamental mathematical models require in depth understanding of the process that requires evaluation of physical properties and quantification of their complex inter-relationships. Since it will not be possible to quantify all these complexities, the accuracy of the fundamental models are constrained with the assumptions made. Thus, fundamental models are not one-to-one representations of the reality; they only signify rational perceptions.

Empirical models are developed by curve fitting to the experimental data. Process variables are selected and the response variables are measured. Uncertainties are taken into consideration by means of a statistically designed experimental plan. The empirical correlation found as a result of this experimental plan has no physical meaning, and it is specific to the particular material, particular geometry and configuration of the equipment, and also the constraints of the equipment related with the boundary conditions (Khalloufi et al. 2009).

Mathematical modeling is a cognitive field that offers practical solutions to complex problems. Thus, fundamental and empirical models are also not as black and white, but there are grey areas in which fundamental concepts are applied and the constants of the fundamental model are evaluated from statistically designed experiments. One such model is the diffusion model.

4.1 The Continuum Model (Diffusion Model)

The diffusion model is a combination of fundamental and empirical approaches. The pore space is treated as continuum consistent with the macroscopic scale appearance in which all microscopic effects are lumped into an effective diffusivity. Effective diffusivity is described in terms of Fick's 2nd law of diffusion:

A. Gueven and Z. Hicsasmaz, *Pore Structure in Food*,
SpringerBriefs in Food, Health, and Nutrition, DOI: 10.1007/978-1-4614-7354-1_4,
© Alper Gueven and Zeynep Hicsasmaz 2013

$$V \frac{\partial \varphi}{\partial t} = \nabla (D_{\text{eff}} \nabla \varphi) \qquad (4.1)$$

where φ is moisture (kg/m^3) or energy $\left(\rho C_p T, \text{ W/m}^3 \right)$, t is the time (s), D_{eff} is the effective moisture (m^2/s), or thermal diffusivity (m^2/s), which can be expressed by:

$$D_{\text{eff}} = \frac{\varepsilon D}{\tau} \qquad (4.2)$$

in which ε is the porosity of the material (dimensionless), D is the actual diffusivity in the pores (m^2/s) and τ is the tortuosity factor (dimensionless). Tortuosity is a correction factor that lengthens the diffusive path accounting for the fact that the pores do not follow straight line paths. This results in a reduction in local diffusion fluxes. The main drawback of the diffusion model is: it assumes that the process is mainly diffusion controlled, and the tortuosity correction may not be able to provide an explanation on other mechanisms involved.

The model ignores the effect of chemical interaction between the solid walls and the fluid phase (moisture or oil). Diffusivities are lumped over the whole porous medium, and thus are independent of local concentrations. However, puffed, baked, extruded, dried, frozen and freeze-dried processed porous solid foods do not exhibit a uniform pore structure, i.e. pore sizes vary within several orders of magnitude (Dogan et al. 2013; Gueven and Hicsasmaz 2011; Kocer et al. 2007; Hicsasmaz et al. 2003; Hicsasmaz and Clayton 1992). This further leads to local variations in cell wall thicknesses (Trater et al. 2005). In addition to that the pore space is formed of three types of pores, namely interconnected pores, dead-end pores and non-interconnected, closed pores (Gueven and Hicsasmaz 2011; Hicsasmaz and Clayton 1992). Therefore, concentrations of transported quantities and their diffusivities are subject to local variations. Initial concentrations are assumed to be uniform within the sample. Instant saturation of the surface is assumed in the case of moisture diffusion, and thus boundary layer resistance is ignored. The geometry is assumed to be a perfect slab, cylinder or sphere. Swelling and shrinkage are not taken into consideration.

4.2 Empirical and Semi-Empirical Models

The most common empirical model used to express moisture diffusion is the Weibull distribution function:

$$\frac{M_t}{M_i} = 1 - \exp\left[-\left(\frac{t}{\alpha} \right)^{\beta} \right] \qquad (4.3)$$

where M_t is the moisture content at time t (kg/kg), M_i is the initial moisture content (kg/kg), t is time (s), and α and β are the model parameters. In the Weibull distribution function, α is the scale parameter and is the reciprocal of the process rate constant, while β determines the shape of the distribution. Weibull model provides information on the kinetics of moisture diffusion (Marabi et al. 2003), but does not provide information on the effects of the pore structure and the transport mechanisms.

4.3 Capillary Models

Experimental studies on porous food matrices suggested that moisture and oil transport in food materials is driven by the capillary pressure (Carbonell et al. 2004) described by Laplace-Young equation:

$$P_c = -\frac{4\sigma \cos \theta}{D} \tag{4.4}$$

where P_c is the capillary pressure (Pa), σ is the surface tension (N/m), θ is the contact angle and D is the capillary diameter (m). The capillary rise, h (m) curve (Aguilera et al. 2004) can be depicted from a force balance known as the Lucas-Washburn equation:

$$h = k\sqrt{t} \tag{4.5}$$

and

$$k = \sqrt{\frac{\sigma D \cos \theta}{4\mu}} \tag{4.6}$$

in which k is the capillary coefficient (m/$\sqrt{s}$), μ is the liquid viscosity (Pa-s) and t is time (s). However, discrepancies from the experimental curve were encountered since the porous food was regarded as a bundle of cylindrical straight capillaries with an effective mean diameter. Carbonell et al. (2004) introduced a tortuosity correction factor (τ^2, dimensionless) in which the capillary coefficient was redefined:

$$k = \sqrt{\frac{\sigma D \cos \theta}{4\tau^2 \mu}} \tag{4.7}$$

4.4 The Pore Network Model

The advantage of the pore network model is: it does not require assumptions and approximations on transport properties such as effective diffusivity or tortuosity. It is a realistic pore structure model that simulates the measured pore structure parameters such as the total pore volume, the non-interconnected pore volume and the pore size distribution. Then, transport problems can further be simulated numerically on the suggested pore network. Conclusions can then be implemented by comparing the simulation results with the experimental data.

The classical capillary model assumes the porous medium to be formed of pores with an effective mean pore size. The network model accounts for the interconnectivity of pore segments with various sizes within the limitations of a lattice geometry. The simplest network is the one in which capillary segments are connected in descending order, i.e., large pore segments are always followed by progressively smaller ones in the direction of intrusion into the network. This is a 1-D network (Fig. 4.1a) called the straight capillary model. The straight capillary

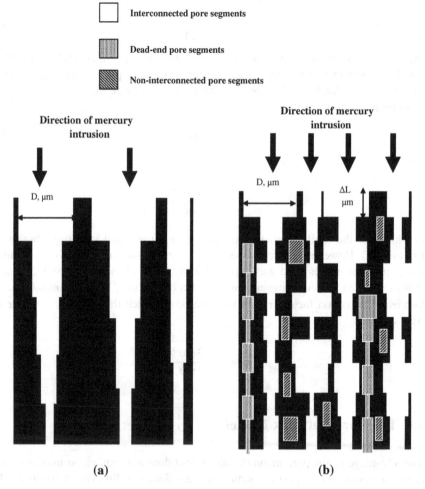

Fig. 4.1 Hypothetical representation of the corrugated pore model: **a** The straight capillary model, **b** The corrugated capillary model

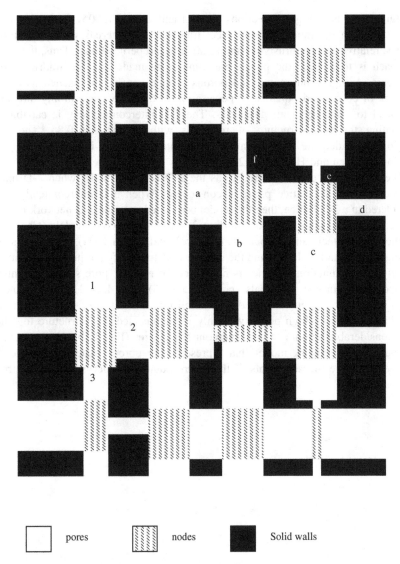

Fig. 4.2 Schematic presentation of the 2-D network model. Surface pores have three neighbors (1–2–3) and interior pore have six neighbors (a–b–c–d–e)

model is used to evaluate pore size distributions from porosimetry data, i.e., mercury intruded pore volume versus capillary pressure.

The corrugated pore concept (Tsetsekou et al. 1991) is an improvement on the straight capillary network. It offers 1-D interconnectivity allowing random distribution of capillary diameters in a parallel bundle of corrugated pores (Fig. 4.1b). This type of network allows integration of neck-and-bulge assemblies into the pore structure. The corrugated pores can be designed as assemblies of cylindrical, triangular, slit or polygonal sections. Segura and Toledo (2005) found that pore shapes used in the model do not affect transport characteristics such as drying curves, relative vapor diffusivity, and relative liquid permeability. Thus, the usual approach is to assume the porous medium to be made up of cylindrical pore segments. A highly interconnected porous sample can be approached as a 1-D network of corrugated pores since penetration of the pore space in any direction will lead to flooding in all directions. The 1-D interconnectivity is capable of simulating the microstructure of materials with porosities ≤ 0.85. The 1-D topology is inadequate in quantitative verification of the non-interconnected pore volume (Gueven and Hicsasmaz 2011).

The network model is referred to as the 2-D network model when 2-D inter-connectivity on the lattice plane is considered. When 3-D interconnectivity is considered in a 3-D lattice, then the model is referred to as the 3-D network model. There are two possible ways of building 2- and 3-D network models (Chan and Hughes 1988; Steele and Nieber 1994; Deepak and Bhatia 1994). One way is not to consider junction volumes and the other way is to specify the junction volumes. Specifying the junction volumes is necessary for realistic pore structure simulations of high porosity materials (porosity > 0.85) (Fig. 4.2). Highly expanded foods such as breads and extruded snacks fall into this category. These materials are characterized by high porosity, highly interconnected pore structure together with considerable amount of non-interconnected pores (Fig. 4.1b). However, it is obvious that a network model that represents the porous material realistically requires sensitive measurements of the macroscopic and microscopic pore structure parameters.

Chapter 5
Simulation of the Pore Structure of Food Materials

Gueven and Hicsasmaz (2011) studied the microstructure of bread and cookie samples based on a 1-D corrugated pore network (Tsetsekou et al. 1991) (Fig. 4.1b). Experimental data on porosity and pore size distribution (Hicsasmaz and Clayton 1992) were used to construct the pore network. Intrusion of non-wetting mercury was simulated on the constructed porous network. The simulation results were compared with the experimental mercury porosimetry data in terms of mercury intruded pore volume versus capillary pressure (Figs. 5.1 and 5.2).

The algorithm (Fig. 5.3) consists of four sections: initiation, distribution of capillary segments, mercury intrusion, evaluation of volume, surface area and pore population distributions based on the diameter of capillary segments. The volume-based distribution $(D_v(D))$ is defined as:

$$D_v(D) = \frac{dV_p}{dD} = \frac{dV_p}{dP_c}\frac{dP_c}{dD} = \frac{P_c}{D}\frac{dV_p}{dP_c} \qquad (5.1)$$

where $D_v(D)$ is evaluated from Laplace-Young equation (Eq. 4.4) in terms of dV_p/dP_c that can be obtained from both experimental and simulated mercury intrusion curves. V_p is the mercury intruded pore volume (μm^3), D is the capillary diameter (μm) and P_c is the capillary pressure (Pa). $D_v(D)$ provides information on the pore volume intruded upon incremental increase in the capillary pressure, and thus it emphasizes the contribution of relatively larger capillary segments to the pore volume. Surface area-based distribution $(D_s(D))$ is defined as:

$$D_s(D) = \frac{dS}{dD} = \frac{dV}{dD}\frac{dS}{dV} = \frac{4}{D}D_v(D) \qquad (5.2)$$

using the cylindrical geometry for the capillary segments. $D_s(D)$ provides information on the surface area of the pores intruded upon incremental increase in the capillary pressure. Volumetric effects are suppressed in $D_s(D)$, and thus it emphasizes the contribution of smaller capillary segments to the pore volume. Population-based distribution $(D_N(D))$ is defined as:

A. Gueven and Z. Hicsasmaz, *Pore Structure in Food*,
SpringerBriefs in Food, Health, and Nutrition, DOI: 10.1007/978-1-4614-7354-1_5,
© Alper Gueven and Zeynep Hicsasmaz 2013

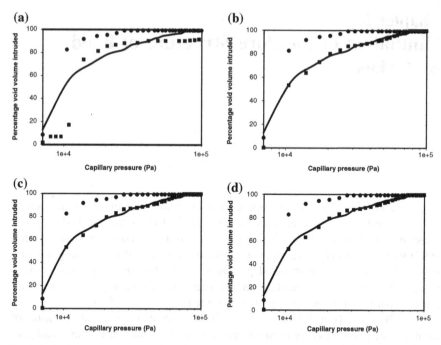

Fig. 5.1 Search for model parameters needed to simulate experimental mercury intrusion curves for the cookie sample (adopted from Gueven and Hicsasmaz 2011): *solid line* experimental, *filled circle* straight capillary model (capillary segments in a unit cell arranged in descending order, **a** *filled square* present model, D_1–D_2 (2 diameter sub-ranges), $\Delta L = 200$ μm, **b** *filled square* present model, D_1–D_3 (3 diameter sub-ranges), $\Delta L = 200$ μm, **c** *filled square* present model, D_1–D_3 (3 diameter sub-ranges), $\Delta L = 150$ μm, **d** *filled square* present model, D_1–D_3 (3 diameter sub-ranges), $\Delta L = 250$ μm

$$D_N(D) = \frac{dN}{dD} = \frac{dN}{dS}\frac{dS}{dD} = \frac{4}{\pi D^2 \Delta L} D_v(D) \tag{5.3}$$

where ΔL is the pore segment length (μm). $D_N(D)$ emphasizes relative populations of pore segments in various size ranges. $D_N(D)$ cannot be calculated from experimental intrusion curves since pore segment length is not a parameter in the straight capillary model. Presence of common peaks in these distributions provides information on the range(s) of dominant capillary size(s).

The model parameters and the experimentally determined pore structure characteristics (Tables 5.1, 5.2 and 5.3) were supplied to the model in the initiation section (Fig. 5.7). The model parameters were: the number of capillary segments, number of diameter sub-ranges and their limits, geometric placement of capillary segments in the unit cell, capillary segment length and the seed number in the random number generation routine. Sample porosity, maximum and minimum pore diameters, mercury intrusion curves, and the pore size histograms were supplied as experimental data (Hicsasmaz and Clayton 1992). The proposed

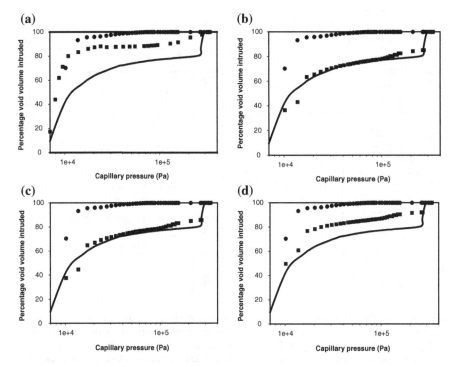

Fig. 5.2 Search for model parameters needed to simulate experimental mercury intrusion curves for the bread sample (adopted from Gueven and Hicsasmaz 2011): *solid line* experimental, *filled circle* straight capillary model (capillary segments in a unit cell arranged in descending order, **a** *filled square* present model, $D_1–D_{12}$, $\Delta L = 100$ μm, **b** *filled square* present model, $D_1–D_{12}$, D_{5min} and D_{6min} after D_1 and D_2, $\Delta L = 100$ μm, **c** *filled square* present model, $D_1–D_{12}$, D_{5min} and D_{6min} after D_1 and D_2, $\Delta L = 80$ μm, **d** *filled square* present model, $D_1–D_{12}$, D_{5min} and D_{6min} after D_1 and D_2, $\Delta L = 200$ μm

network was based on the assumption that there is at least 1 μm distance between two randomly-placed, maximum diameter pore neighbors. This assumption is the backbone of the 1-D network which does not allow any adjacent pores to touch, intersect or overlap.

Experimental pore structure data impose a primary set of constraints on the model, and that is why the pore network simulation is inherent to the sample. The capillary segments are distributed in the second section of the algorithm to obtain the experimentally measured porosity within the limitations of the pore size histogram characteristic to the sample (Table 5.1). Random distribution of the capillary segments was done by using a random number generation routine. 1-D network models in the literature (Steele and Nieber 1994; Bryant et al. 1993; Ionnides and Chatzis 1993) have been suggested for relatively low porosity materials compared with foods. Simulation of the pore structure of food materials with porosities > 0.4 requires extra constraints on the random distribution of capillaries with various sizes (Gueven and Hicsasmaz 2011). For this purpose, the

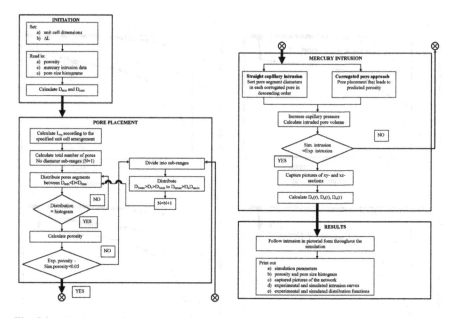

Fig. 5.3 Algorithm used to simulate the microstructure of the bread and cookie samples. Input experimental data (Hicsasmaz and Clayton 1992) to the algorithm in the initiation section belongs to the real bread or the cookie sample. Results of the decision boxes are evaluated in terms of these data characteristic to the bread or the cookie sample (adopted from Gueven and Hicsasmaz 2011)

Table 5.1 Comparison of experimental cell size histograms with that obtained from the proposed model (Basis for Figs. 5.1 and 5.2)

Cookie sample			Bread sample		
Diamater range (µm)	% of pores within the range		Diamater range (µm)	% of pores within the range	
	Experimental	Simulated		Experimental	Simulated
$D > 140$	12.5	11.8	$D > 180$	7.5	6.9
$140 > D > 120$	12.5	11.9	$180 > D > 160$	2.5	2.5
$120 > D > 110$	6.0	6.2	$160 > D > 140$	7.5	0.2
$110 > D > 100$	6.0	6.1	$140 > D > 100$	12.5	9.3
$100 > D > 70$	12.5	12.5	$100 > D > 80$	7.5	0.4
$70 > D > 60$	12.5	12.3	$80 > D > 60$	15.0	2.1
$60 > D > 55$	27.0	27.0	$60 > D > 40$	2.5	12.1
$55 > D > 20$	6.0	6.5	$40 > D > 20$	12.5	42.0
$D < 20$	6.0	5.7	$D < 20$	22.5	24.5

Adopted from Gueven and Hicsasmaz (2011)

Table 5.2 Effect of simulation parameters on predicted porosity of the cookie sample

Unit cell dimensions ($L_{cell} \times W_{cell} \times H_{cell}$, µm³)	Type of unit cell	Unit square length (L_{sq}, µm)	Diameter sub-ranges (µm)	Pore segment length (ΔL, µm)	Seed number	Predicted porosity
10000 × 10000 × 10000	I	215	$214 \geq D \geq 6$	200	Random	0.40
10000 × 10000 × 10000	II	266	$214 \geq D_1 \geq 160$	200	Random	0.52
10000 × 10000 × 10000	II	266	$160 > D_2 \geq 6$	200	50	0.52
10000 × 10000 × 10000	II	266		250	50	0.52
12500 × 12500 × 6400	II	266		200	50	0.52
10000 × 10000 × 7500	II	266		100	50	0.52
10000 × 10000 × 7500	II	266	$214 \geq D_1 \geq 160$	150	Random	0.53
10000 × 10000 × 7500	II	266	$160 > D_2 \geq 20$	200	Random	0.53
10000 × 10000 × 7500	II	266	$20 > D_2 \geq 6$	250	Random	0.53

Adopted from Gueven and Hicsasmaz (2011)

experimental pore size range is divided into a number of diameter sub-ranges (Tables 5.2 and 5.3). The capillary diameters in each sub-range are distributed randomly. Constraints on capillaries that belong to various sub-ranges neighboring each other are imposed as secondary constraints so that the simulated network conforms to the primary constraints dictated by the experimental results.

Gueven and Hicsasmaz (2011) found that porosities ≥ 0.50 characteristic to food materials could be simulated with a unit cell arrangement in which the pore size range was divided into several sub-ranges, and more than a single capillary was accommodated in a unit square (Fig. 5.4). The pore structure of the cookie sample with porosity = 0.54 was successfully simulated with respect to pore size distributions (Table 5.1) and mercury intrusion curves (Fig. 5.1) using 6 pore diameter sub-ranges, while 12 diameter sub-ranges was required for the bread sample (Table 5.1; Fig. 5.2). Maximum and minimum pore diameters were accommodated next to each other to allow more pore space (Fig. 5.4).

Basic calculation steps start by assigning a unit square length that allows accommodation of maximum and minimum pores adjacently such that (Fig. 5.4):

$$L_{sq} = \frac{D_{max} + D_{min} + 2\,\mu m}{\sqrt{2}} \tag{5.4}$$

Then, the number of corrugated pores on the x–y plane, N_p, is calculated as:

$$N_p = \left(\frac{L_{cell}}{L_{sq}}\right)^2 \times n_{sq} \tag{5.5}$$

where L_{cell} is the length of the unit cell (µm), L_{sq} is the unit square length (µm), n_{sq} is the number of capillaries in a unit square. The number of capillary segments in a corrugated pore, N, is calculated as:

Table 5.3 Effect of simulation parameters on predicted porosity of the bread sample

Unit cell dimensions ($L_{cell} \times W_{cell} \times H_{cell}$, μm³)	Type of unit cell	Unit square length (L_{sq}, μm)	Diameter sub-ranges (μm)	Pore segment length (ΔL, μm)	Seed number	Predicted porosity
10000 × 10000 × 10000	I	221	$220 \geq D \geq 5$	100	Random	0.39
17250 × 17250 × 3400	I	221		100	Random	0.41
17250 × 17250 × 3400	III	232	$220 \geq D_1 \geq 165$ $160 \geq D_2 \geq 5$ $65 \geq D_3,D_4 \geq 5$ $32 \geq D_5,D_6 \geq 5$ $66 \geq D_7 \geq 5$ $99 \geq D_8,D_9 \geq 5$	100	Random	0.74
17250 × 17250 × 3400	III	232		80	Random	0.85
17250 × 17250 × 3400	III	232	$220 \geq D_1 \geq 165$ $160 \geq D_2 \geq 5$	100	Random	0.85
17250 × 17250 × 3400	III	232	$65 \geq D_3,D_4 \geq 5$ $32 \geq D_5,D_6 \geq 5$	100	50	0.85
17250 × 17250 × 3400	III	232	$66 \geq D_7 \geq 5$ $99 \geq D_8,D_9 \geq 5$	200	50	0.85
17250 × 17250 × 2400	III	232	$110 \geq D_{10},D_{11} \geq 5$ $77 \geq D_{12} \geq 5$	80	Random	0.85
13800 × 13800 × 5250	III	232		100	Random	0.84
17250 × 17250 × 3400	III	232	$220 \geq D_1 \geq 165$ $160 \geq D_2 \geq 5$ $65 \geq D_3,D_4 \geq 5$	80	Random	0.85
17250 × 17250 × 3400	III	232	$D_{5min},D_{6min} = 5$ $32 \geq D_5,D_6 > 5$ $66 \geq D_7 \geq 5$	100	Random	0.85
17250 × 17250 × 3400	III	232	$99 \geq D_8,D_9 \geq 5$ $110 \geq D_{10},D_{11} \geq 5$ $77 \geq D_{12} \geq 5$	200	Random	0.85

Adopted from Gueven and Hicsasmaz (2011)

$$N = \frac{H_{cell}}{\Delta L} \tag{5.6}$$

where H_{cell} is the unit cell thickness (μm), and ΔL is the pore segment length (μm). Total number of capillary segments, N_T, in the unit cell is then:

$$N_T = N_p \times N \tag{5.7}$$

Then, the placement of capillary segments has to be performed, first by distributing the whole diameter range, $D_{min} \leq D \leq D_{max}$ to build the network. Placement of capillary segments is monitored by labeling each capillary segment in a corrugated pore, m, as (j_m, k_m, l_m) where j, k and l signify the x-, y-, and z-coordinates of the center of any capillary in the corrugated pores. Capillary segments with diameters between $D_{min} \leq D \leq D_{max}$ were distributed using randomly generated integers such that N_p number of corrugated pores (Eq. 5.5) contained N number of capillary

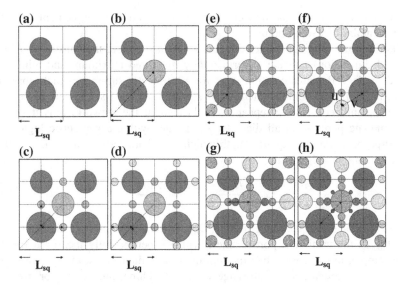

Fig. 5.4 Placement of capillary segments in the unit cell with 12 diameter sub-ranges (D_1-D_{12}) (adopted from Gueven and Hicsasmaz 2011). **a** D_1 pores, **b** D_2 pores, **c** D_3 pores, D_4 pores, **d** D_5 pores, D_6 pores, **e** D_7 pores, **f** D_8 pores, D_9 pores, **g** D_{10} pores, D_{11} pores, **h** D_{12} pores $u = \frac{L_{sq}}{2} - \frac{D_{4min}}{2}$ $v^2 = \left(\frac{L_{sq}}{2}\right)^2 + \left(\frac{L_{sq}}{2} - \frac{D_{8max}}{2}\right)^2 = \left(\frac{D_{1min}}{2} + \frac{D_{8max}}{2}\right)^2$

segments (Eq. 5.6). The resulting distribution of capillary segment diameters is compared with the experimental cell size histograms (Table 5.1) as the first constraint (Fig. 5.3). Capillary segments are redistributed until the capillary segment distribution was 99 % in agreement with the experimental distribution.

Then, the porosity of the simulated network was calculated as:

$$\varepsilon = \frac{\sum_{m=1}^{n} V_m}{V_{cell}} \qquad (5.8)$$

where ε is the porosity (dimensionless), V_m is the volume of each capillary segment (μm^3) and V_{cell} is the unit cell volume (μm^3). Experimentally measured porosity is the second constraint (Fig. 5.3). If network porosity < experimentally measured porosity, more capillary segments are incorporated into the network using more diameter sub-ranges. For this purpose, the whole diameter range, $D_{min} \leq D \leq D_{max}$, is first divided into two sub-ranges in which both sub-ranges are represented in a unit square to simulate first the experimental capillary size distribution within 99 % agreement with the experimentally measured pore diameter distribution. Network porosity (Eq. 5.8) is again compared with the experimentally measured porosity. The number of diameter sub-ranges in a unit square is increased until the network porosity is within 95 % agreement with the experimentally measured porosity.

Diameter sub-ranging (Tables 5.2 and 5.3) is done under the constraints of the unit cell dimensions and the experimentally measured maximum and minimum pore diameters. Basic calculation steps for 12 diameter sub-ranges that represent the bread sample are demonstrated in this section. The key issue in pore placement is that the unit cell should repeat itself throughout the network, thus it should accommodate pores from all diameter sub-ranges within a geometric logic. For this purpose, first a unit square length which is a characteristic of the unit cell is defined such that (Fig. 5.4):

$$L_{sq} = \frac{D_{max} + D_{min} + 2\,\mu m}{\sqrt{2}} \tag{5.9}$$

where L_{sq} is the unit square length (μm), D_{max} is the maximum pore diameter (μm) and D_{min} is the minimum pore diameter (μm) experimentally measured for the bread sample, in this case. Then the minimum pore diameter in the D_1 sub-range that can accommodate a D_2 sub-range without touching, intersecting or overlapping is calculated (Fig. 5.4) as:

$$D_{1_{min}} = L_{sq}\sqrt{2} - D_{2_{max}} \quad \text{where} \quad D_{1_{min}} < D_{max} \quad \text{and} \quad D_{min} < D_2 < D_{max} \tag{5.10}$$

D_3 and D_4 sub-ranges are distributed within the space left from D_1 and D_2 sub-ranges. In fact, D_3 and D_4 pore segments belong to the same diameter sub-range with the difference that the D_3 pore segments lie parallel to the x-axis and the D_4 pore segments lie parallel to the y-axis such that (Fig. 5.4):

$$\frac{D_{3_{max}}\,(\text{or }D_{4_{max}})}{2} = \frac{L_{sq}}{2} - \frac{D_{1_{min}}}{2} \quad \text{where} \quad D_{1_{min}} < D_{max} \quad \text{and} \quad D_{min} < D_3 < D_{3max} \tag{5.11}$$

D_3 and D_4 sub-ranges will touch, intersect or overlap with D_1 and D_2 sub-ranges unless $D_{3_{max}}$ (or $D_{4_{max}}$) $< D_{1_{min}}$. D_3/D_4 sub-range should be constrained such that:

$$\frac{D_{2_{max}}}{2} + \frac{D_{3_{max}}\,(\text{or }D_{4_{max}})}{2} \leq \frac{L_{sq}}{2} \tag{5.12}$$

An iterative procedure on $D_{1_{min}}$ is performed to recalculate $D_{2_{max}}$ and $D_{3_{max}}$ (Eqs. 5.10 and 5.11) until the condition given by Eq. (5.12) is fulfilled setting the sub-range limits as: $D_{1_{min}} \leq D_1 \leq D_{max}$, $D_{min} < D_2 < D_{2_{max}}$, $D_{min} < D_3$ (or D_4) $< D_{3_{max}}$. D_5 and D_6 sub-ranges (D_5 pore segments lie parallel to the x-axis, D_6 pore segments lie parallel to the y-axis) are placed next to the D_1 sub-range (Fig. 5.4) as:

$$D_{5_{max}}\,(\text{or }D_{6_{max}}) = \frac{L_{sq}}{2} - \frac{D_{1_{min}}}{2} \quad \text{where} \quad D_{min} < D_5\,(\text{or }D_6) < D_{5_{max}} \tag{5.13}$$

Then, D_7 sub-range is placed at the corners (Fig. 5.4) as:

$$D_{7_{max}} = \frac{L_{sq}\sqrt{2}}{2} - \frac{D_{1_{min}}}{2} \quad \text{where} \quad D_{min} < D_7 < D_{7_{max}} \tag{5.14}$$

Next, D_8/D_9 sub-range (D_8 pore segments lie parallel to the x-axis, D_9 pore segments lie parallel to the y-axis) are placed on the sides adjacent to the D_3/D_4 sub-range (Fig. 5.4) such that the condition that yields a smaller numerical value in between:

$$D_{8_{max}} \text{ (or } D_{9_{max}}) = \frac{L_{sq} - D_{3_{min}}}{2} \quad \text{or} \quad D_{8_{max}} \text{ (or } D_{9_{max}}) = \frac{2L_{sq}^2 - D_{1_{min}}^2}{2\left(L_{sq} + D_{1_{min}}\right)} \tag{5.15}$$

is chosen not to allow D_8/D_9 sub-range to intersect, touch or overlap with the D_1 or D_3/D_4 sub-ranges. The sub-range limits are: $D_{min} < D_8 (\text{or } D_9) < D_{8_{max}}$. D_{10}/D_{11} sub-range (D_{10} pore segments lie parallel to the x-axis, D_{11} pore segments lie parallel to the y-axis) is placed similar to the D_3/D_4 sub-range in between the D_1, D_2 and D_3/D_4 sub-ranges. No touching, intersection or overlapping of the D_{10}/D_{11} sub-ranges with the D_3/D_4 sub-ranges requires:

$$D_{10_{max}} \text{ (or } D_{11_{max}}) = \frac{L_{sq}}{2} - \frac{D_{2_{min}}}{2} - \frac{D_{3_{min}}}{2} \tag{5.16}$$

Intersection of the D_{10}/D_{11} sub-range with the D_1 sub-range can be prevented by finding the larger of the two D_1 neighbors of a D_{10} capillary segment (Fig. 5.4). For this purpose, x-coordinate for the center of a D_{10} capillary segment ($x_{D_{10}}$) and its radius, and the x-coordinate for the center of the larger D_1 capillary segment (x_{D_1}, y_{D_1}) are substituted into the equation of a circle and a y-coordinate is calculated:

$$y = y_{D_1} + \sqrt{\left(\frac{D_1}{2}\right)^2 - (x_{D_{10}} - x_{D_1})^2} \tag{5.17}$$

Then the following condition is checked:

$$y > y_{D_{10}} - \frac{D_{10}}{2} \quad \text{for} \quad D_1(j_n, k_n, l_n) \quad \text{and} \quad y < y_{D_{10}} + \frac{D_{10}}{2} \quad \text{for} \quad D_1(j_n, k_{n+1}, l_n) \tag{5.18}$$

If the above condition is satisfied, D_1 and D_{10}/D_{11} are expected to touch, intersect or overlap, so an iterative procedure has to be applied on $D_{10_{max}}$ until Eq. (5.18) is reversed. For the D_{11} sub-range, an x-coordinate should be calculated from Eq. (5.17) by interchanging x's with y's. This is due to the geometric orientation of the D_{11} sub-range. Then, the conditions given for the y-coordinate of the D_{10} sub-range (Eq. 5.18) should be calculated for the x-coordinate of the D_{11} sub-range, and compared with capillary segments $D_1(j_n, k_n, l_n)$ and $D_1(j_{n+1}, k_n, l_n)$. The limits for D_{10}/D_{11} capillary segments are then determined as: $D_{min} < D_{10} < D_{10_{max}}$

and $D_{min} < D_{11} < D_{11_{max}}$. D_{12} sub-range is distributed within the space left between the D_1 and D_2 sub-ranges such that (Fig. 5.4):

$$D_{12_{max}} = \frac{L_{sq}\sqrt{2} - D_{2_{min}} - D_{3_{min}}(\text{or } D_{4_{min}})}{2} \quad \text{where} \quad D_{min} < D_{12} < D_{12_{max}} \quad (5.19)$$

In this analysis sub-ranges D_1–D_9 belong to the interconnected pore space, while sub-ranges D_{10}–D_{12} belong to the non-interconnected pore space.

The third section of the algorithm is simulation of the mercury intrusion curve (Fig. 5.3). Simulation of mercury intrusion is constrained (third constraint) by the experimental mercury intrusion curve characteristic to the sample (Figs. 4.1 and 4.2). Pressure on the network with acceptable porosity is incrementally increased (Eq. 4.4), and the mercury-intruded volume was followed assuming cylindrical capillary segments such that

$$V_p = \sum_{i=1}^{k} V_{p_i} \quad (5.20)$$

where V_p is the mercury-intruded capillary volume (μm^3) and V_{p_i} is the volume of capillaries intruded (μm^3) upon incremental rise in pressure. If the distribution of capillary segments successful in simulating porosity fails in fulfilling the constraint on experimental mercury intrusion curves, capillary segments are redistributed by increasing the number of diameter sub-ranges and/or by applying extra constraints on the placement of various pore diameters (Gueven and Hicsasmaz 2011). The last step of the simulation is the calculation of the volume-, surface area- and number-based distribution functions (Eqs. 5.1–5.3).

A maximum porosity of 0.52 was obtained (141 330 capillary segments, for the cookie sample) when two diameter sub-ranges were constrained to neighbor each other (Table 5.3). The most effective model parameter on the accuracy of porosity simulation was the number of diameter sub-ranges, which directly influenced the number of corrugated pores in the unit cell. Porosity of the network increased as the number of diameter sub-ranges increased, thus imposing additional constraints on the model. Nine diameter sub-ranges led to a porosity of 0.74, and the maximum porosity that could be obtained using a 1-D pore network was 0.85 (4,560,959 capillary segments, for the bread sample) (Table 5.3). These results showed that a 1-D network model constrained with an experimentally determined pore size histogram and porosity can readily simulate the micro-structure of the real sample for porosity values ≤ 0.85 (Figs. 4.2 and 5.1). Thus, application of the network model can overcome the need to define an average pore size related effective diffusivity in studying transport problems through porous food materials.

The constraint on the pore size histogram had to be relaxed for the sample with porosity > 0.85 (the bread sample), and porosity values > 0.85 could not be obtained from the 1-D geometric network model. Also, predictions on the non-interconnected pore volume were inaccurate (Gueven and Hicsasmaz 2011). Use of mercury intrusion curves refined the proposed network. The 1-D network with three diameter sub-ranges was capable of simulating the pore structure of the cookie sample very close to reality in comparison with the experimental pore size histogram, porosity and the mercury intrusion curve (Fig. 5.2). Thus, a 1-D network refined by any experimental data on transport phenomena can reveal the microstructure of the food material under consideration. Therefore, this network model offers a first step to simulate other transport phenomena relevant to the food material in question.

Preliminary work on a 2-D network model with specified junction volumes (Fig. 4.2) proved out to be promising in simulating porosity values >0.85. Results of the 2-D network were more accurate with respect to the experimental pore size histograms, experimentally measured porosity and experimental mercury intrusion curves (unpublished). Preliminary studies showed that the 2-D approach also improved the predictions on the non-interconnected pore volume, but the model still needs to be refined to have better predictions.

Chapter 6
Applications of the Pore Network Model to Food Systems

6.1 Moisture Diffusion

Although the pore network model is widely used in petroleum reservoir engineering, soil science and catalysis, very few applications of the model to food systems are available. One of the applications is the 1-D geometric network applied to capillary penetration as described above (Gueven and Hicsasmaz 2011). Besides, 2-D geometric network models to simulate drying (Surasani et al. 2009) and moisture diffusion (Pratomak et al. 2010) were proposed.

Prakotmak et al. (2010) studied moisture diffusion through the model food system of banana foam obtained upon drying of banana puree. The proposed 2-D network was based on pore size distributions obtained from SEM micrographs of the foam. The pores were assumed to have cylindrical geometry and the pore segment length was assumed constant as in the case of Gueven and Hicsasmaz 2011. Pore junctions had a connectivity of 4 (Fig. 4.2). The simulated sample had the same cross-sectional area with the sample used in the experiments (Prakotmak et al. 2010). Moisture adsorption between the surrounding air and the banana foam took place under isothermal conditions, thus the diffusion coefficient was constant. Diffusion of moisture adsorbed on the food surface through the pore network was simulated by integrating the constructed pore network with Fick's Second Law of Diffusion (Eq. 4.1) where φ is moisture content (kg/kg db). No moisture was transferred to the bottom surface of the banana foam, since it was placed on an opaque glass dish. The length and width of the dried banana foam was about 11 times of its thickness and thus, the banana foam was assumed to be an infinite slab. Accordingly, moisture moved along the material thickness. Movement of moisture through the pores was simulated by solving Eq. (4.1) through the constructed network using a finite element solution. The initial condition for the finite element solution was uniform moisture content at the foam surface. The results of simulation were validated by comparing the simulated and experimental moisture sorption curves.

Pratomak et al. (2010) concluded that the resistance to moisture diffusion decreased with increase in porosity, thus became vulnerable to moisture diffusion.

A. Gueven and Z. Hicsasmaz, *Pore Structure in Food*,
SpringerBriefs in Food, Health, and Nutrition, DOI: 10.1007/978-1-4614-7354-1_6,
© Alper Gueven and Zeynep Hicsasmaz 2013

The dried banana foam system was identified by very large pores interconnected towards the surface which acted as a very efficient medium for moisture diffusion into the interior. On the contrary, rapid moisture diffusion also means higher rates of drying. If surface pores were smaller forming a skin as in the case of baked and extruded foods moisture sorption could be slower due to the skin barrier. Prachayawarakorn et al. (2008) who studied the effect of pore assembly on the rate of drying, found drying rates were higher when the large pores are located at the surface and the smaller pores are located in the interior.

6.2 Drying

Surasani et al. (2009) approached the drying problem as an interconnected pore-and-throat assembly in a 2-D network (Fig. 4.2). The 2-D pore-and-throat assembly was constructed on a rectangular cell (Segura 2007). A random number generation routine, as well as statistical distributions such as a log-normal distribution (Segura 2007) can be used for random distribution of the pore segments. Length of the throats and center-to-center distance between the pores are kept constant. The network porosity has to be in accordance with the measured porosity of the specimen.

The network was saturated with water at constant temperature. Bone dry air blew over the network at a higher temperature. All the moisture that can move through the network towards the surface due to the vapor pressure gradient will be evaporated. Then, the falling rate period begins which is identified by a continuous water vapor phase and discontinuous clusters of liquid in the pore network. During the falling rate period, liquid flow, evaporation, and vapor diffusion are simultaneous, and are influenced by capillary forces and gravity at the solid—gas boundary. In partially filled throats, the meniscus moves as evaporated liquid diffuses through the air, and at the same time, liquid flows in either direction depending on the boundary liquid pressure at the meniscus. Vapor is transferred due to molecular diffusion through the network. Liquid in the pores move depending on the average saturation of connected throats. Liquid can be pumped from the throat with the highest liquid pressure to all other meniscus throats. Further, drying rates are lower under the effect of gravity. Gravity counteracts capillary forces. Thus capillary pumping is limited to a finite distance resulting in a stabilized drying front which recedes into the network as drying proceeds. In later stages, the surface throats dry out, and lose their connections with the interior liquid clusters. This causes a sharp decrease in the rate of drying followed by the breakthrough of the gas phase to the bottom of the network.

Chapter 7
Conclusions

As a result, it can be concluded that a pore network model is capable of simulating the internal pore structure of porous food samples with relatively higher porosity (porosity >0.5). Realistic network simulations were possible for materials with porosity <0.85 using the 1-D network under the constraints of experimental pore size histograms, experimentally measured porosity and the experimental mercury intrusion curves. The major difference between the proposed network model and the ones applied for lower porosity materials is that a complete random distribution of all capillary sizes could not be used. Instead, the capillary diameter range was divided into a number of diameter sub-ranges each of which was separately distributed. Also, the network model had to be further constrained with respect to various diameter sub-ranges being placed next to each other.

The mathematics became more tedious as the complexity of the model increased. 3-D reconstructions from 2-D image analysis using quantitative stereology also require tedious mathematical calculations, and the methodology needs to be standardized and validated. The geometric network model of pore structure is a powerful mathematical model that can be integrated with X-ray microtomographic measurements, especially to simulate thermal conductivity which is very difficult to measure in porous food materials due to the order of magnitude differences in the thermal conductivity of air cells and thin cell walls. The pore network model and 3-D reconstructed images can be integrated to predict the average thermal conductivity through porous food materials, as well as local variations in thermal conductivity throughout the material. Fourier's Law of Heat Conduction can be used to simulate conduction heat transfer through the network using the composite treatment of series and parallel resistances.

A. Gueven and Z. Hicsasmaz, *Pore Structure in Food*,
SpringerBriefs in Food, Health, and Nutrition, DOI: 10.1007/978-1-4614-7354-1_7,

References

Agbisit, R., Alavi, S., Cheng, E., Herald, T., & Trater, A. (2007). Relationships between microstructure and mechanical properties of cellular cornstarch extrudates. *Journal of Texture Studies, 38*, 199–219.

Aguilera, J. M., Michel, M., & Mayor, G. (2004). Fat migration in chocolate: Diffusion or capillary flow in a particulate solid?—a hypothesis paper. *Journal of Food Science, 69*, R167–R174.

Ahrné, L., Andersson, C. G., Floberg, P., Rosén, J., & Lingnert, H. (2007). Effect of crust temperature and water content on acrylamide formation during baking of white bread: steam and falling temperature baking. *LWT—Food Science and Technology, 40*, 1708–1715.

Altamirano-Fortoul, R., Le-Bail, A., Chevallier, S., & Rosell, C. M. (2012). Effect of the amount of steam during baking on bread crust features and water diffusion. *Journal of Food Engineering, 108*, 128–134.

Autio, K., & Salmenkallio-Martilla, M. (2001). Light microscopic investigations of cereal grains, doughs and breads. *Lebensmittel-Wissenschaft und Technologie, 34*, 18–22.

Babin, P., Della Valle, G., Chiron, H., Cloetens, P., Hoszowska, J., & Pernot, P. (2006). Fast X-ray tomography analysis of bubble growth and foam setting during breadmaking. *Journal of Cereal Science, 43*, 393–397.

Bai, Y., Rahman, M. S., Perera, C. O., Smith, B. G., & Melton, L. D. (2002). Structural changes in apple rings during convection airdrying with controlled temperature and humidity. *Journal of Agricultural and Food Chemistry, 50*, 3179–3185.

Baik, O. D., & Marcotte, M. (2002). Modeling the moisture diffusivity in a baking cake. *Journal of Food Engineering, 56*, 27–36.

Brun, F., Mancini, L., Kasae, P., Favretto, S., Dreossi, D., & Tromba, G. (2010). Pore3D: A software library for quantitative analysis of porous media. *Nuclear Instruments and Methods in Physics Research A, 615*, 326–332.

Bryant, S. L., Mellar, D. W., & Cade, C. A. (1993). Physically representative network models of transport in porous media. *American Institute of Chemical Engineers Journal, 39*(3), 387–396.

Campbell, G. M., & Mougeot, E. (1999). Creation and characterization of aerated food products. *Trends in Food Science & Technology, 10*, 283–296.

Carbonell, S., Hey, M. J., Mitchell, J. R., Roberts, C. J., Hipkiss, J., & Vercauteren, J. (2004). Capillary flow and rheology measurements on chocolate crum/sunflower oil mixtures. *Journal of Food Science, 69*, E465–E470.

Carson, J. K., Lovatt, S. J., Tanner, D. J., & Cleland, A. C. (2004). Experimental measurements of the effective thermal conductivity of a pseudo-porous food analogue over a range of porosities and mean pore sizes. *Journal of Food Engineering, 63*, 87–95.

Chan, D. Y. C., & Hughes, B. D. (1988). Simulating flow in porous media. *Physics Review: Series A, 38*(8), 4106–4120.

A. Gueven and Z. Hicsasmaz, *Pore Structure in Food*,
SpringerBriefs in Food, Health, and Nutrition, DOI: 10.1007/978-1-4614-7354-1,
© Alper Gueven and Zeynep Hicsasmaz 2013

Chanvrier, H., Appelqvist, I. A. M., Bird, A. R., Gilbertt, E., Htoon, A., Li, Z., et al. (2007). Processing of novel elevated amylose wheats: Functional properties and starch digestibility of extruded products. *Journal of Agricultural and Food Chemistry, 55*, 10248–10257.

Cheng, E. M., Alavi, S., Pearson, T., & Agbisit, R. (2007). Mechanical-acoustic and sensory evaluations of cornstarch-whey protein isolate extrudates. *Journal of Texture Studies, 38*, 473–498.

Cho, K. Y., & Rizvi, S. S. H. (2009). 3D Microstructure of supercritical fluid extrudates. II: Cell anisotropy and the mechanical properties. *Food Research International, 42*, 603–611.

Datta, A. K. (2007). Porous media approaches to studying simultaneous heat and mass transfer in food processes. I: Problem formulations. *Journal of Food Engineering, 80*, 80–95.

Datta, A. K., Sahin, S., Sumnu, G., & Keskin, S. Ö. (2007). Porous Media Characterization of breads baked using novel heating modes. *Journal of Food Engineering, 79*, 106–116.

Deepak, P. D., & Bhatia, S. K. (1994). Transport in capillary network models of porous media: theory and simulation. *Chemical Engineering Science, 49*(2), 245–257.

Dogan, H., Gueven, A., & Hicsasmaz, Z. (2013). Extrusion cooking of lentil flour (*lens culinaris – red*)—corn starch—corn oil mixtures. *International Journal of Food Properties, 16*, 341–358.

Fang, Q., & Hanna, F. M. (2000). Mechanical properties of starch-based foams as affected by ingredient formulations and foam physical characteristics. *Transactions of the ASAE, 43*, 1715–1723.

Frisullo, P., Barnabà, M., Navarini, L., & Del Nobile, M. A. (2012). Coffea arabica beans microstructural changes induced by roasting: An X-ray microtomographic investigation. *Journal of Food Engineering, 108*, 232–237.

Gogoi, B. K., Alavi, S. H., & Rizvi, S. S. H. (2000). Mechanical properties of protein-stabilized starch-based supercritical fluid extrudates. *International Journal of Food Properties, 3*(1), 37–58.

Gropper, M., Moraru, C., & Kokini, J. L. (2002). Effect of specific mechanical energy on properties of extruded protein–starch mixtures. *Cereal Chemistry, 79*, 429–433.

Guessasma, S., Chaunier, L., Della Valle, G., & Lourdin, D. (2011). Mechanical modelling of cereal solid foods. *Trends in Food Science & Technology, 22*, 142–153.

Gueven, A., & Hicsasmaz, Z. (2011). Geometric network simulation of high porosity foods. *Applied Mathematical Modelling, 35*, 4824–4840.

Haedelt, J., Pyle, D. L., Beckett, S. T., & Niranjan, K. (2005). Vacuum induced bubble formation in liquid-tempered chocolate. *Journal of Food Science, 70*, E159–E164.

Hamdami, N., Monteau, J. Y., & Le Bail, A. (2003). Effective thermal conductive evolution as a function of temperature and humidity during freezing of high porosity model. *Transactions of the Institution of Chemical Engineers, Part A, 81*, 1123–1128.

Hicsasmaz, Z., & Clayton, J. T. (1992). Characterization of the pore structure of starch based food materials. *Food Structure, 11*(2), 115–132.

Hicsasmaz, Z., Yazgan, Y., Bozoglu, F., & Katnas, S. (2003). Effect of Polydextrose-substitution on the cell structure of the high-ratio cake system. *LWT-Lebensmittel Wissenschaft and Technologie, 36*, 441–450.

Horigane, A. K., Naito, S., Kurimoto, M., Irie, K., Yamada, M., & Motoi, H. (2006). Moisture Distribution and Difusion in Cooked Spaghetti Studied by NMR Imaging and Diffusion Model. *Cereal Chemistry, 83*, 235–242.

Hussain, M. A., Rahman, M. S., & Ng, C. W. (2002). Prediction of pores (porosity) in foods during drying: generic models by the use of hybrid neural network. *Journal of Food Engineering, 51*, 239–248.

Ioannidis, M. A., & Chatzis, I. (1993). Network modeling of pore structure and transport in porous media. *Chemical Engineering Science, 48*(5), 951–972.

Ishida, N., Takano, H., Naito, S., Isobe, S., Uemura, K., Haishi, T., et al. (2001). Architecture of baked breads depicted by a magnetic resonance imaging. *Magnetic Resonance Imaging, 19*, 867–874.

James, B. (2009). Advances in "wet" electron microscopy techniques and their application to the study of food structure. *Trends in Food Science & Technology, 20*, 114–124.

Karkle, E. L., Alavi, S., & Dogan, H. (2012). Cellular architecture and its relationship with mechanical properties in expanded extrudates containing apple pomace. *Food Research International, 46*, 10–21.

Khalloufi, S., Rivera, C. A., & Bongers, P. (2009). A theoretical model and its experimental validation to predict the porosity as a function of shrinkage and collapse phenomena during drying. *Food Research International, 42*, 1122–1130.

Khalloufi, S., Rivera, C. A., & Bongers, P. (2010). A fundamental approach and its experimental validation to simulate density as a function of moisture content during drying processes. *Journal of Food Engineering, 97*, 177–187.

Kocer, D., Hicsasmaz, Z., Bayındirli, A., & Katnas, S. (2007). Cell structure of the high-ratio cake with polydextrose as a sugar- and fat-replacer. *Journal of Food Engineering, 78*, 953–964.

Lagrain, B., Boeckx, L., Wilderjans, E., Delcour, J. A., & Lauriks, W. (2006). Non-contact Ultrasound Characterization of Bread Crumb: Application of the Biot–Allard Model. *Food Research International, 39*, 1067–1075.

Le-Bail, A., Dessev, T., Leray, D., Lucas, T., Mariani, S., Mottollese, G., et al. (2011). Influence of the amount of steaming during baking on the kinetics of heating and on selected quality attributes of crispy rolls. *Journal of Food Engineering, 105*, 379–385.

Leonard, A., Blacher, S., Nimmol, C., & Devahastin, S. (2008). Effect of far-infrared radiation assisted drying on microstructure of banana slices: An illustrative use of X-ray microtomography in microstructural evaluation of a food product. *Journal of Food Engineering, 85*, 154–162.

Li, K., Gao, X.-L., & Subhash, G. (2006). Effects of cell shape and strut cross-sectional area variations on the elastic properties of three-dimensional open-cell foams. *Journal of the Mechanics and Physics of Solids, 54*, 783–806.

Lim, K. S., & Barigou, M. (2004). X-ray micro-computed tomography of cellular food products. *Food Research International, 37*, 1001–1012.

Liu, Z., & Scanlon, M. G. (2003). Predicting mechanical properties of bread crumb. *Transactions of I Chem E, 81*, 212–220.

Luyten, H., Plijter, J. J., & van Vliet, T. (2004). Crispy/crunchy crust of cellular solids food: a literature review with discussion. *Journal of Texture Studies, 35*, 445–492.

Madioulia, J., Sghaiera, J., Lecomteb, D., & Sammoudaa, H. (2012). Determination of porosity change from shrinkage curves during drying of food material. *Food and Bioproducts Processing, 90*, 43–51.

Marabi, A., Livings, S., Jacobson, M., & Saguy, I. (2003). Normalized weibull distribution for modeling rehydration of food particulates. *European Food Research and Technology, 217*, 311–318.

Mayor, L., & Sereno, A. M. (2004). Modelling shrinkage during convective drying of food material: a review. *Journal of Food Engineering, 61*, 373–386.

Mesa, N. J. E., Alavi, S., Singh, N., Shi, Y. C., Dogan, H., & Sang, Y. (2009). Soy protein-fortified expanded extrudates: Baseline study using normal corn starch. *Journal of Food Engineering, 90*, 262–270.

Moraru, C. I., & Kokini, J. L. (2003). Nucleation and expansion during extrusion and microwave heating of cereal foods. *Comprehensive Reviews in Food Science and Food Safety, 2*, 120–138.

Olurin, O. B., Arnold, M., Korner, C., & Singer, R. F. (2002). The investigation of morphometric parameters of aluminum foams using micro-computed tomography. *Materials Science and Engineering, 328*, 334–343.

Pai, D. A., Blake, O. A., Hamaker, B. R., & Campanella, O. H. (2009). Importance of extensional rheological properties of fiber-enriched corn extrudates. *Journal of Cereal Science, 50*, 227–234.

Parada, J., Aguilera, J. M., & Brennan, C. (2011). Effect of guar gum content on some physical and nutritional properties of extruded products. *Journal of Food Engineering, 103*, 324–332.

Pontente, H., Ernst, W., & Oblotzki, J. (2006). Description of the foaming process during the extrusion of foams based on renewable resources. *Journal of Cellular Plastics, 42*, 241–253.

Prachayawarakorn, S., Prakotmak, P., & Sopornronnarit, S. (2008). Effect of pore size distribution and pore-architecture assembly on drying characteristics of pore networks. *International Journal of Heat and Mass Transfer, 51*, 344–352.

Prakotmak, P., Soponronnari, S., & Prachayawarakorn, S. (2010). Modelling of moisture diffusion in pores of banana foam mat using a 2-D stochastic pore network: Determination of moisture diffusion coefficient during adsorption process. *Journal of Food Engineering, 96*, 119–126.

Primo-Martín, C., van Dalen, G., Meinders, M. B. J., Don, A., Hamer, R. H., & van Vliet, T. (2010). Bread crispness and morphology can be controlled by proving conditions. *Food Research International, 43*, 207–217.

Rahman, M. S. (2003). A theoretical model to predict the formation of pores in foods during drying. *International Journal of Food Properties, 6*, 61–72.

Robin, F., Engmann, J., Pineau, N., Chanvrier, H., Bovet, N., & Della Valle, G. (2010). Extrusion, structure and mechanical properties of complex starchy foams. *Journal of Food Engineering, 98*, 19–27.

Roca, E., Guillard, V., Guilbert, S., & Gontard, N. (2006). Moisture migration in a cereal composite food at high water activity: Effects of initial porosity and fat content. *Journal of Cereal Science, 43*, 144–151.

Sablani, S. S., & Rahman, M. S. (2003). Using neural networks to predict thermal conductivity of food as afunction of moisture content, temperature and apparent porosity. *Food Research International, 36*, 617–623.

Saguy, Í. S., Marabi, A., & Wallach, R. (2005). Liquid imbibition during rehydration of dry porous foods. *Innovative Food Science and Emerging Technologies, 6*, 37–43.

Scanlon, M. G., & Zghal, M. C. (2001). Bread properties and crumstructure. *Food Research International, 34*, 841–864.

Segura, L. A. (2007). Modeling at pore-scale isothermal drying of porous materials: Liquid and vapor diffusivity. *Drying Technology, 25*, 1677–1686.

Segura, L. A., & Toledo, P. G. (2005). Pore-level modeling of isothermal drying of pore networks: Effects of gravity and pore shape and size distributions on saturation and transport parameters. *Chemical Engineering Journal, 111*, 237–252.

Stasiak, M. W., & Jamroz, J. (2009). Specific surface area and porosity of starch extrudates determined from nitrogen adsorption data. *Journal of Food Engineering, 93*, 379–385.

Steele, D. D., & Nieber, J. L. (1994). Network modeling of diffusion coefficients for porous media. I. Theory and model development. II. Simulations. *Soil Science Society American Journal, 58*(5), 1337–1345.

Stokes, D. J., & Donald, A. M. (2000). In situ mechanical testing of dry and hydrated breadcrumb in the Environmental Scanning Electron Microscope (ESEM). *Journal of Materials Science, 35*, 599–607.

Surasani, V. K., Metzger, T., & Tsotsas, E. (2009). A non-isothermal pore network drying model with gravity effect. *Transport in Porous Media, 80*, 431–439.

Trater, A. M., Alavi, S., & Rizvi, S. S. H. (2005). Use of non-invasive X-ray microtomography for characterizing microstructure of extruded biopolymer foams. *Food Research International, 38*, 709–719.

Troutman, M. Y., Mastikhin, I. V., Balcom, B. J., Eads, T. M., & Ziegler, G. R. (2001). Moisture migration in soft-panned confections during engrossing and aging as observed by magnetic resonance imaging. *Journal of Food Engineering, 48*, 257–267.

Troygot, O., Saguy, I. S., & Wallach, R. (2011). Modeling rehydration of porous food materials: I. Determination of characteristic curve from water sorption isotherms. *Journal of Food Engineering, 105*, 408–415.

Tsetsekou, A., Androutsopoulos, G. P., & Mann, R. (1991). Mercury porosimetry hysteresis and entrapment predictions based on a corrugated random pore model. *Chemical Engineering Communications, 110*, 1–29.

van Dalen, G., Blonk, H., van Haalts, H., & Hendriks, C. L. (2003). 3-D imaging of foods using X-ray microtomography. *G.I.T. Imaging and Microscopy, 3*, 18–21.

Wagner, M., Quellec, S., Trystram, G., & Lucas, T. (2008). MRI evaluation of local expansion in bread crumb during baking. *Journal of Cereal Science, 48*, 213–223.

Index

A
Anisotropy, 4, 14, 20
Average pore size, 1, 17, 21, 27, 36, 39

B
Baking, 2, 3
Brittleness, 18
Bulk density, 14

C
Capillary flow, 2, 7
Capillary penetration, 2, 7, 8, 39
Capillary pressure, 3, 7, 23, 26–28
Capillary segment, 25, 27–29, 32–36
Cell shape, 20
Cell structure, 1, 3, 7
Cell wall thickness, 1, 3, 5, 7, 8, 12, 14, 17, 20, 21, 27, 39
Cellular foods, 17, 18, 20
Composite foods, 1, 2
Compression test, 17, 18
Compressive strength, 1, 17, 21, 27, 39
Continuum model, 21
Convenience foods, 1, 2
Corrugated pore, 26, 27, 32, 36
Crispiness, 2, 3, 18

D
Destructive imaging technique, 7, 8, 11
Drying, 1- 3, 9, 13, 21, 39, 40

E
Effective diffusivity, 8, 22, 24, 36
Elastic recovery, 14

Empirical model, 21, 22
Engineering strain, 17
Engineering stress, 17
Environmental scanning microscopy (ESEM), 7, 9

F
Failure stress, 17
Fick's Second Law of Diffusion, 2, 39
Flexural modulus, 19
Flexure strength, 19
Flexure test, 17, 18
Fracture stress, 17
Freezing, 2- 4
Frying, 2
Fundamental model, 21

G
Gibson & Ashby model, 4, 19, 20

I
Image analysis, 7, 8, 10, 11, 13- 15, 41
Interconnected pore volume, 36
Interconnectivity, 1, 3, 5, 7, 12, 14, 17, 20, 21, 26, 27, 39

J
Jaggedness analysis, 18

L
Laplace-Young equation, 7, 23, 27
Light microscopy, 7- 9, 13
Liquid extrusion porosimetry, 7, 8

A. Gueven and Z. Hicsasmaz, *Pore Structure in Food*,
SpringerBriefs in Food, Health, and Nutrition, DOI: 10.1007/978-1-4614-7354-1,
© Alper Gueven and Zeynep Hicsasmaz 2013